D0218536

CHRIS SPRING & JULIE HUDSON

silk in africa

THE BRITISH MUSEUM PRESS

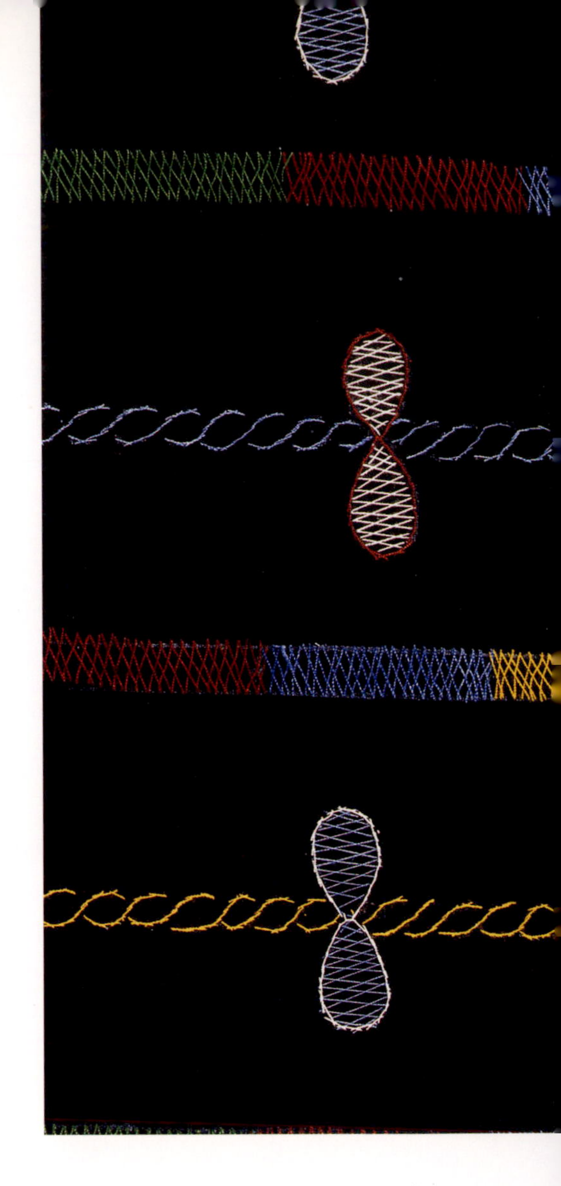

First published in 2002 by The British Museum Press
A division of The British Museum Company Ltd
46 Bloomsbury Street, London WC1B 3QQ

A catalogue record for this book is available from the British Library

ISBN 0-7141-2563-6

Commissioning editor: Suzannah Dick
Designer: Paul Welti
Editor: Caroline Brooke Johnson
Photographer: Mike Rowe
Cartographer: Olive Pearson
Origination in Hong Kong by AGP
Printing and binding in Hong Kong by C & C Offset

COVER: Detail from a silk textile (*lamba*); Betsileo people, Madagascar. (See pp.72–3)

INSIDE COVER: Detail from a ceremonial cloth (*itagbe*); Nigeria. (See pp.70–1)

PAGE ONE: Detail from a silk and cotton wrap; Asante people, Ghana. (See pp.60–61)

PREVIOUS PAGES: Detail from a hanging screen; Ethiopia. (See pp.48–9)

THESE PAGES: Detail from an indigo-dyed cotton cloth with silk embroidery; Senegal or Gambia.

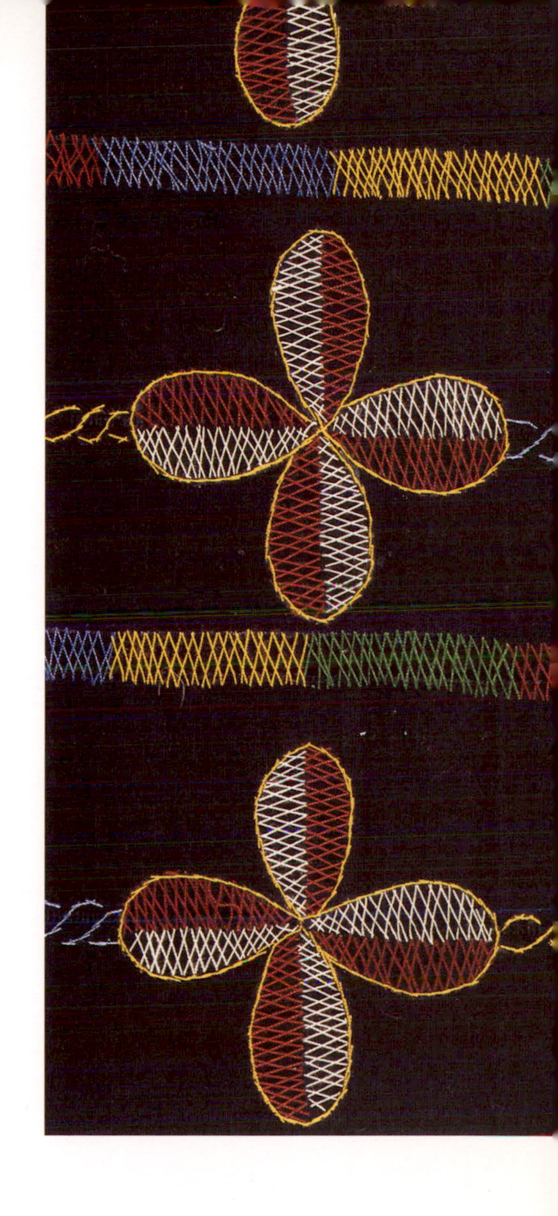

contents

THIS PAGE: Shoes; Hausa people, Nigeria.
OPPOSITE PAGE: Details from a selection of textiles:
TOP: Silk textile; Betsileo people, Madagascar. (See pp.72–3)
LEFT: Silk and cotton wrap; Asante people, Ghana. (See pp.60–1)
RIGHT: Embroidered tunic; Central and Northern Ethiopia. (See pp.52–3)
BELOW: Ceremonial cloth (*Itagbe*); Nigeria. (See pp.70–1)

introduction

Silk is a prestigious material, often used to produce textiles and clothing associated with rank, wealth and social status. In Africa silk is produced and used less extensively than cotton and wool – both geographically and socially. The textiles in this book come from North Africa (Morocco, Algeria, Tunisia, Libya and Egypt), Ethiopia, West Africa (mainly Ghana and Nigeria) and Madagascar.

Some of the earliest written records of indigenous silk production in Africa come from Tunisia; the tenth-century geographer Ibn Hawqal refers to the flourishing industry

in Gabès. Mulberry trees were cultivated suggesting that the *Bombyx mori*, literally 'mulberry cocoon', silkworm was being raised there. This variety was also reared in Madagascar – during the reign of Radama I (1817–25) local cultivation was encouraged and mulberry trees from Mauritius were introduced. Until then Chinese silk, derived from the *Bombyx mori*, had been imported by Arab and Indian traders. As in Tunisia, raw silk was probably obtained by heating the cocoons to soften the gum binding the filaments, and reeling (unwinding) the threads directly from two or more cocoons at a time to form a continuous even strand.

As well as cultivated production several varieties of wild silk are known in Africa. In the highlands of Madagascar the indigenous *Borocera Madagascariensis* produces a coarse greyish-brown silk which resists dye. It is esteemed for its durability and used almost exclusively for weaving burial shrouds. The savanna region of northern Nigeria is home to two varieties of wild silk collected from the cocoons of the *Anaphe infracta* and *Anaphe moloneyi* genus moths, which breed mainly on tamarind trees. Unlike *Bombyx* varieties, the *Anaphe infracta* caterpillars form clusters of cocoons with a communal brown silk outer casing. These communal pods are boiled in an alkaline solution to separate the silk filaments from the gummy substances that bind the outer casing together. After several rinses, the silk is dried and the fibres are spun. The yarn is coarse, greyish-brown and lacking in lustre. The *Anaphe moloneyi* caterpillar produces clusters of whitish cocoons unprotected by a casing. The light-beige silk yarn is spun directly from these cocoons and is mainly used for embroidery.

Among the Betsileo people of Madagascar, the caterpillars are raised in containers filled with mulberry leaves at the weaving workshop.

IMPORT AND TRADE

The trans-Saharan trade routes provided a vital source of silk yarn and manufactured silk cloths. Magenta waste silk, known in Nigeria as *alharini* (derived from *harir*, the Arabic word for silk), originally came from Tunisia as a by-product of the local silk industry. Later unwrought silk from French and Italian mills was imported into Gabès and Tunis and subsequently traded to Kano. The mid-nineteenth century German traveller, Heinrich Barth, recorded that 300–400 camel loads of unwrought silk were traded from Ghadamis in Libya to Kano annually.

The Asante people of Ghana exploited other sources to obtain silk. By the 1730s

Samples of red-dyed spun silk thread and unspun green-dyed silk yarn imported to Kano via the trans-Saharan trade routes.

manufactured silk or composite cloths, acquired from Europe as diplomatic gifts or via the West African coastal trade, were being unravelled for their yarn. This practice continued until the early twentieth century. Similarly, silk weavers in Ethiopia relied on imported silk thread and fabrics from India, China and Arabia.

NORTH AFRICA

Following the Arab invasion of the seventh century, weaving became a male profession and horizontal ground looms were introduced. *Tiraz* factories producing luxurious silk garments for the rulers and the Muslim élite were established. Textiles were included as tribute items or presented as gifts and rewards, particularly under the hedonistic Fatimids (909–1171) in Egypt and Tunisia. By the twelfth century specialist silk weaving guilds had been set up in Tunis. Production also flourished in Fès in Morocco which boasted 467 inns for merchants and over 3,000 houses of *tiraz*. Later, under the Marinids (1269–1465) this industry reportedly supported 20,000 artisans.

Weaving

Silk weaving was largely based in the towns and cities of North Africa, rather than in rural areas. Treadle looms of various types were used from the sixteenth century. Perhaps the most complex was the large draw loom, used in Fès to produce intricately patterned silk textiles and sumptuous ceremonial belts (*hizam*), often further embellished with gold (see pp.34–7). The decoration and the manufacture of these belts mirrored the complex dissemination of weaving techniques and ideas, as successive waves of refugees introduced technical and stylistic innovations. Many Andalucian exiles, including those of Jewish ancestry, settled in North African towns and cities following the fall of Granada in 1492. In the town of Mahdia today, small draw looms operated by two people, are used to weave silk ribbons (*hashiya*) with complex central motifs. This skill was introduced to Mahdia by Jewish weavers from Tripoli in Libya towards the end of the nineteenth century. Such ribbons are used in Tunisia and Libya to form curtains and hangings, or to decorate marriage costumes (see pp.38–9).

Flat-woven silk textiles, often incorporating simple stripes or checks, were

Weaving the ceremonial woman's garment (*rida' ahmar*) using a treadle loom in Mahdia, Tunisia.

made on hand-operated treadle looms in Libya, Tunisia and Egypt. The practice of inserting complex tapestry-woven designs using shuttle boats is still seen today on Jerba island in the production of *biskri* (wrap-around cloths) and in Mahdia for the prestigious *rida' ahmar* (red outer garment), both of which are used by women at marriage. Tapestry-woven cloths also come from the town of Naqada in the Nile Valley, Egypt. Prestigious items of men's dress such as the *'aba'*, a sleeveless outer robe worn throughout the Middle East, and women's modesty garments of silk and cotton were tapestry-woven in Naqada (see pp.26–7).

The introduction of the flying shuttle to treadle looms, which the weaver operates by pulling on a short cord, allowed broader cloths to be created and increased output. On Jerba island and in Mahdia today this technique is used to create silk cloths with elaborate geometric motifs by inserting numerous supplementary shed sticks through the warp. These subtle designs, often produced in two-tone, are known as *khwatim trabelsi* (Tripolitanian seal). The name reveals the origin of this cloth: before the arrival of Jewish weavers from Libya, such textiles were imported from Jerba island's neighbour, Tripolitania.

Urban embroidery

While silk weaving in North Africa is essentially a male activity, embroidery is a female art. Although there are clear distinctions between urban and rural embroidery, both types are produced mainly for domestic use. Quite distinctive regional embroidery styles and techniques evolved, reflecting different external influences.

Amongst the finest embroideries are those worked on linen from Algiers; they reached the height of their elegance during the Ottoman period (sixteenth–eighteenth centuries). Beautiful hangings and curtains were produced by women for the bridal trousseau. These were usually made in panels separated by brocade ribbons and covered with exuberant floral and vegetal designs. Wealthy women also prepared personal items such as the *beniqa* (bath cap) which was used at the *hammam* to wrap damp hair. At home women wore reversible shawls (*tanshifa*) draped around the shoulders or covering the hair. Reserved for use on ceremonial occasions, they were extensively embroidered with foliate and floral designs inspired by Italian Renaissance or Turkish patterns (see pp.30–1).

A more formal embroidery style was adopted in Fès for cloths designed as home furnishings. Worked in monochrome on a cotton base cloth, the repeating patterns are arranged in bands: stylized versions of trees, hands or birds feature along the borders (see pp.32–3). Cloths for home furnishing from the town of Azemmour are decorated with stylized female figures with raised arms, birds flanking a central vase and mythical beasts, recalling pattern elements derived from Renaissance Italy and Spain.

Embroidery class at a school in the rue Marey in Algiers during the 1940s. Pupils were taught embroidery skills by French nuns who themselves had learnt local techniques from Algerian families.

Rural embroidery and patterns

In contrast to the formal composition of urban embroidery, rural patterns reflect concerns about fertility, generate community well-being and protect against the evil eye. Embroidery communicates cultural beliefs in an exuberant and engaging manner. Striking solar, lunar and pillar motifs, worked in brilliantly coloured silks on heavy wool, characterize the spectacular wrap-around cloths (*hram nfasiy*) worn by Tunisian women of El-Jem. Scattered in a seemingly random manner alongside these central designs are appealing camels, fish, hands, scorpions and brides with upraised hands. Worn for the first time during wedding ceremonies, these patterns protect against the evil eye. Solar-inspired designs on the black and white wedding dresses from Siwa Oasis in Egypt are also thought to have protective qualities. Linear embroidery creates a 'sunburst' pattern entirely covering the front of the dress (see pp.24–5). Today plastic amulets in the form of hands, fish and horseshoes act as protection. Woollen head-shawls from the

Gabès area of southern Tunisia have human and animal motifs, as well as palm trees and moons worked in slightly uneven chain or herringbone stitch. These patterns echo the smaller designs on the *hram nfasiy*, suggesting a sub-Saharan inspiration.

ETHIOPIA

Silk has played an important part in the social and religious life of Ethiopia from the earliest days of the kingdom of Aksum, which was converted to Christianity in the fourth century. It was imported in large quantities from India, Arabia and China, and stored in vast caverns in the central highlands of Ethiopia. One of the hereditary titles of the governor of Shoa province was 'Keeper of the Silk Caves'. From these storehouses Ethiopian emperors would make prodigious gifts of silk to other churches in Christendom. Apart from some isolated, historical traditions such as on Pate island off the Kenyan coast, Ethiopia has long been the only major silk weaving region in eastern Africa.

Silk and the Church

Silk was used to make ceremonial umbrellas, to bind sacred books, to cover wooden altar tablets and, most spectacular of all, to weave the imposing hangings which were an essential feature of Ethiopian churches. One of these hangings survives in the collections of The British Museum, and the history and significance of silk in Ethiopian life and culture is woven into the fabric of this single cloth (see pp.48–9). It was designed as the central section of a triptych which would have screened the inner sanctum (*maqdas*) from the main body of an Ethiopian Orthodox Christian church. The practice of adorning the interiors of churches with silk hangings was widespread in the Byzantine empire – a tradition that survived longer in Ethiopia than it did in the rest of Christendom.

The mid-eighteenth century curtain in the British Museum's collection is the largest tablet-woven textile in the world, and though a profoundly Christian artefact, it was probably created by a guild of Muslim or Jewish (Falasha) weavers in the city of Gondär. It is woven entirely of imported Chinese silk, and the figures that appear on it are depicted in such detail that the soldiers can be seen to be carrying firearms of Indian make. The event commemorated is probably the lying-in-state of King Bakaffa who died in 1730. He, his wife Queen Mentuab and their young son Iyasu are all depicted wearing the *matab* (plaited band of blue silk), which was a symbol of their Christian faith.

Silk and the State

During the nineteenth century a succession of enlightened rulers began to unify the Christian empire of Ethiopia. In common with the Asante of Ghana and the Merina of Madagascar, the complex hierarchy of religious, secular and military officialdom in this empire needed delineation. An impressive range of woven or embroidered silk costumes was created to define the status of the wearer.

The *shamma*

One such textile was the *shamma*, a toga-like cotton shawl worn by both men and women throughout Ethiopia (see pp.46–7). In the mid-nineteenth century only a very few were permitted to wear the *margaf*, a type of *shamma* that boasted a decorative band (*tibeb*) with supplementary weft patterns in red, yellow and blue silk. In common with the *kente* cloth of the Asante, Ethiopian weavers first obtained their supplies of silk yarn from unravelled imported textiles. Also like the Asante *kente* and the *lamba akotofahana* of the Merina of Madagascar, the silk *tibeb* of the *shamma* became more various in both pattern and colour towards the end of the nineteenth century as it came to represent an increasingly complex social hierarchy. Not only was information about the status of the wearer conveyed by the pattern of the *shamma*, but also by the way in which the garment was draped or folded. This remains a feature of the *shamma* to this day.

The *lamd*

An honorific lion-skin cape was gradually replaced during the nineteenth century by a *lamd* (tailored garment) which was worn by clerics, nobles and high-ranking military and secular officials (see pp.50–1). Guilds of Armenian craftsmen were responsible for the rich silk embroidery, as well as the silver and

A Muslim weaver from Aksum in northern Ethiopia, inserting supplementary shed sticks into the warp to produce a complex pattern on the silk band (*tibeb*) which is incorporated into the *shamma*.

gold filigree with which these capes were embellished. The base cloth and linings were made with imported velvet and silk.

The *kamis*

'A fine lady has a ... shirt [which] is made probably of calico from Manchester, and richly embroidered in silk of divers and various patterns around the neck, down the front, and on the cuffs,' observed Mansfield Parkyns in the mid-nineteenth century.

Ethiopian noblewomen embroidered their costume in this way, using Chinese silk yarn on a base cloth of calico imported from Manchester (see pp.52–3). Today similar dresses are worn by a much larger number of women, although they remain a symbol of status (see pp.54–5). The base cloth is now the finest hand-woven Ethiopian cotton; this is usually embroidered by a Muslim man, often using rayon and lurex rather than silk.

While there is a large Muslim population in Ethiopia, as well as followers of other faiths, this book is concerned with the predominantly Christian northern and central highlands where silk is particularly significant. The ancient emirate of Harär close to the border with Somalia in the east is an exception. With its strong Muslim population and culture, it was renowned for its textiles, many of them silk, including elaborately embroidered wedding shirts.

WEST AFRICA – NIGERIA

The British Museum houses an important collection of textiles and raw materials from Egga (modern Eggan) in northern Nigeria acquired by the Niger Expedition in 1841. Many of the cloths and samples incorporate both locally produced silk and imported European magenta waste silk (*alharini*). Silk weaving was a specialist activity in Egga and surrounding Nupe towns, and the goods were made for export. Distinctive cloths, such as the popular 'guineafowl' design (see pp.68–9), were renowned as far away as Timbuktu.

Cloth types

Male weavers using double-heddle treadle looms produce narrow strips of cloth, which are sewn together to make men's gowns and trousers and women's wraps. Yoruba, Nupe and Hausa weavers are most closely associated with the production of silk or cotton and silk strips. Among the most prestigious cloths are those made for men's robes that use the local *Anaphe infracta* silk, usually with cotton. These beige-coloured cloths are known as *sanyan* in Yoruba and *tsamiya* in Hausa. The 'guineafowl'-patterned cloth originally included a good proportion of *sanyan* silk in its composition. Robes incorporating this cloth were especially appropriate dress at funerals and other traditional ceremonies. Indigenous silk was sometimes dyed with camwood or indigo and combined with imported magenta waste silk for use in gowns or as lining materials.

Embroidered gowns

In early Islamic society gifts of clothing were presented to officials or members of the court to reward service and loyalty. Similarly, in the early nineteenth century elaborately embroidered men's gowns became popular among the ruling Muslim élite of the northern Nigerian emirates. The gowns were a symbol of political and religious affiliation. One of the finest examples of these robes of honour was presented to a senior British officer by the King of Dahomey in about 1862 (see pp.64–7).

Pattern

Northern Nigerian gowns are characterized by their formal composition and the consistency of the patterns used to embellish them (see pp.64–7). Several different specialized artisans are responsible for the separate stages or elements of manufacture. Notably, Islamic scholars draw the preliminary designs on the cloth for the embroidery. The most distinctive design motifs are named 'two knives' (*aska biyu*) and 'eight knives' (*aska takwas*). Their form – long triangular elements arranged in groups of two and three – and their name suggest a protective function. In North Africa swords and daggers are often used as amulets.

GHANA

Two major weaving groups – the Asante and the Ewe people – dominate silk textile production in Ghana. As in Nigeria, men weave narrow strip cloths on horizontal treadle looms. However, Asante and Ewe textiles are distinguished by their use of colour, their sophisticated composition and, most notably, by their complex weft-float patterning (see pp.60–3). Imported silk yarns or cloths for unravelling were available from at least the eighteenth century via northern and coastal trade markets.

Asante silk weaving

Two of the most prestigious types of regalia associated with the Asante royal court were gold artefacts and silk textiles. Bonwire, a village close to Kumasi, was established as the major centre of silk weaving. The *Asantehene* (the king) commissioned specific cloths for his exclusive use and maintained a monopoly on certain patterns. Many of the silk cloths were distributed by him to favoured officials.

Patterns and names

Early cotton cloths with simple, solid colour weft blocks in silk were known as *bankuo*. Gradually more complex finely banded weft blocks (*babadua*) were introduced, and were eventually combined with intricate weft-float designs (*adwin*), which incorporated geometric patterns such as triangles, zigzags and lozenges. The finest examples of such cloths were known as *nsaduaso*. Each type of silk cloth was individually named according to its warp patterns, suggesting that the names were selected prior to the introduction of the weft-float patterns. There appears to be no particular logic to the use of names: some

The design on this Asante textile is called 'Fathia Nkrumah' commemorating the Egyptian wife of the former president Kwame Nkrumah, Ghana, Bonwire.

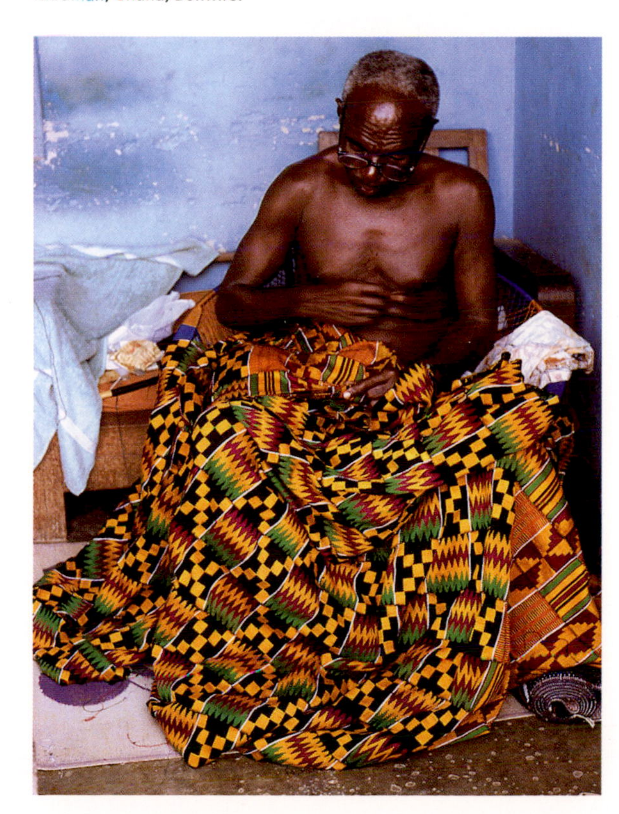

refer to proverbs, such as 'money attracts many relatives', others are descriptive, while some may have historical or personal significance.

In addition to these cloths, sumptuous silk cloths (*asasia*), were made exclusively for the Asantehene. These cloths are distinguished by the use of three instead of two sets of heddles, producing textiles with more intricate patterns and a distinctive twill effect. The names of these cloths are derived from the dominant weft-inlay designs, because the warp is totally concealed.

Ewe silk weaving

Historically the Ewe people did not have a strong centralized government or court, unlike the Asante, so there was no established system of patronage or monopoly of silk cloth production. Ewe cloths could be acquired by anyone with sufficient financial means and were commissioned for specific occasions such as the celebration of marriage, the birth of a child, funerals or a special festival.

The most prestigious type of Ewe cloth is the *adanudo* which can be woven of high-quality cotton but is more usually produced in silk or a silk and rayon mix. It is specially commissioned and in terms of social significance it relates to the Asante *nsaduaso* cloths. Ewe *adanudo* cloths also used weft-float designs which are sometimes geometric but more often representational. Animals, human figures and, recently, inanimate objects derived from daily life, such as stools, forks and knives, feature prominently. On older cloths the patterns are more stylized and less readily identifiable while more recent examples are more naturalistic and explicit in their use of images. Some *adanudo* cloths commemorate specific historic events, such as Ghana's independence in 1957 or the visit of Queen Elizabeth II in 1959. Proverbs provide a rich source of visual imagery often relating directly to everyday experiences.

MADAGASCAR

Madagascar's many distinctively different textile traditions reflect the numerous waves of people who have settled on the island over the centuries. The most prestigious of these textiles have been made of silk, whether from the local variety of silkworm, *landibe* (the big worm), or the more recently introduced Chinese silkworm *Bombyx mori,* known in Madagascar as *landikely* (the little worm). Just as silk in Ethiopia was associated with the emperor and with Christianity, in Madagascar it was associated with royalty, immortality and the deity, 'Fragrant Lord' being a literal translation of one of the Malagasy words for silk. The many different words that describe the variety of silk types and their different qualities indicate the importance of silk in Malagasy culture.

The Merina and Betsileo people are the principal silk weavers in Madagascar, and until recently all weaving, including the production of silk cloth, was undertaken by women. One of the most important woven items in Madagascar is the *lamba*. The term *lamba* is a generic name for all cloth in Madagascar; however, it has come to describe the shawl that is habitually worn by Malagasy men and women, rich and poor.

Silk for the ancestors

Shrouds (*lamba mena*) are the most prestigious – and expensive – of all Malagasy textiles. Cloth woven from silk produced by the local variety of silkworm, though dull in colour, is extremely tough and durable, making it the most appropriate dress in which to approach the afterlife.

From the nineteenth century the royal court and aristocracy of the ruling Merina people of the central highlands began to use elaborately patterned silk cloths called *lamba akotofahana* (see pp.74–7). These textiles could be worn as a mark of status both for the living and, ultimately, for the dead. They were used as an additional shroud in the elaborate 'second-burial' rituals (*famadihana*) which, like much else in Malagasy culture, has its origins in Indonesia. These ceremonies are still widely practised. They are often precipitated by a medium with whom the spirit of the deceased has communicated, perhaps indicating that he, the spirit, is getting chilly in the family tomb. In turn this suggests that the *lamba mena* in which he is wrapped is beginning to deteriorate and needs replacing. During the resulting ceremonies, the deceased is exhumed and wrapped in a new shroud and perhaps an additional shroud of patterned white silk. Other ancestors may also be honoured and celebrated in a similar way. If they have not been interred in the family tomb, their memorials made in stone or wood are wrapped in silk cloth as if these were their mortal remains.

Pattern and colour

Most Malagasy textiles express their predominant design in the warp (see pp.72–81). However, weft-float patterns became an increasingly important feature of Merina textiles in the nineteenth century (see pp.74–7). Merina weavers were able to weave bands of complex pattern because of two principal innovations. Firstly, in the nineteenth century King Radama I introduced the Chinese silkworm to the island, which provided Merina weavers with the long, fine silk threads needed to create weft-float patterns. Secondly, Merina weavers adapted their looms, which are very similar to the technically simple ground looms of North Africa and the Middle East, to include one or more supplementary heddles suspended above the loom. Each of these heddles could then be manipulated to produce a band of complex pattern floating across the warp of the textile (see pp.74–7). In addition, weavers worked with many sheets of paper beside them; each sheet was a written record of the sequence of heddle manipulations required to produce a desired pattern.

Many, if not all, the pattern elements on the *lamba akotofana* were named. Those that

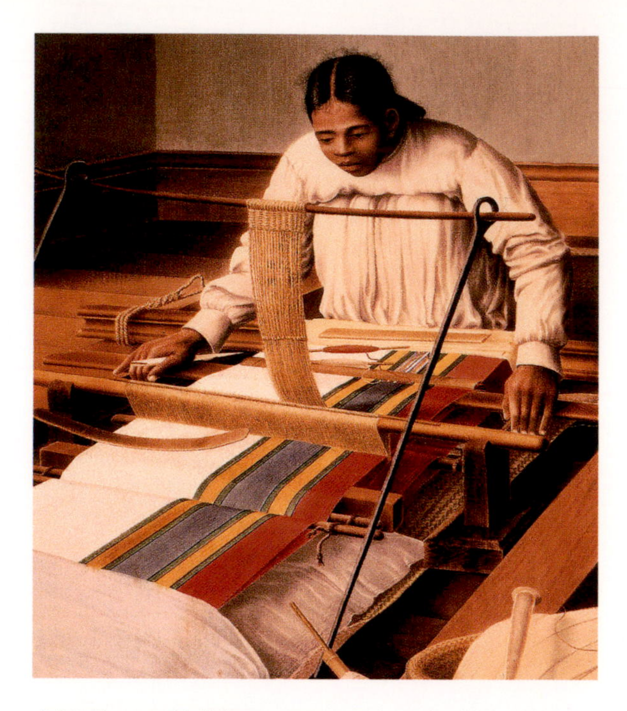

Painted in 1907, this Malagasy weaver uses a single additional heddle suspended high above the warp to produce the bands of complex pattern characteristic of the multicoloured *lamba akotofahana*.

are still recorded include *ravim-pibasy* (a leaf), *vato peratra* (stone shapes arranged as a diamond) and *kintana* (a star). Some patterns are derived from trees and flowers – these probably relate to the system of honours that helped to define the complex hierarchy of the Merina kingdom in the nineteenth century. The literal translation of the Malagasy word for 'to have honours' (*manamboninahitra*) is 'he who has the flower of grass'. Just as the complexity of the designs of the Ethiopian *shamma* were an indication of the wearer's status, the combination of warp stripes and float-patterned bands on the *lamba akotofahana* fulfilled a similar purpose. A distinction was also made between *amboradara lehibe* (the great brocade), and *amboradara kely* (the small brocade) as it was between the silver brocades and the gold brocades.

Colours undoubtedly added a further dimension to the significance of the cloth. However, Malagasy colour symbolism is extremely subtle. For example, green is associated with mourning, and the term *lamba maitso*, literally 'green cloth', is one of the terms used to describe mourning cloth, though green may not be its actual colour. Similarly *lamba mena*, literally 'red cloth', is a term used to describe the shrouds used for burying the dead, though they are not necessarily red in colour. A seventeenth-century account of Malagasy textiles refers to red-dyed silk cloth being used to 'wrap distinguished corpses in', implying the *lamba mena* were once predominantly red. The colour red had been associated for centuries with royalty (who among the Merina were thought to be immortal), vitality and mystical power. The cloth became so closely identified with what red represented that when the colours changed, the name did not. A similar transformation may have taken place with the *rida' ahmar*, the most prestigious of the silk textiles in Mahdia, Tunisia. Its name again literally translates as 'red shawl', but apart from the complex and multicoloured float patterns, it is mainly black.

During the colonial period the Malagasy monarchy and the Merina kingdom lost its power and authority, and at the same time the significance of the colour and patterning of the *lamba akotofahana* dwindled. In the nineteenth century white had been a colour associated with subordinate people, commoners and slaves, but in the post-colonial period the wearing of white silk *lamba* has became a mark of prestige and status. Some of the patterns of the *lamba akotofahana*, as well as the technique of weaving them, were also preserved, though in a white-on-white design (see pp.78–9). An innovative group of weavers from the town of Arivonimamo, near the capital Antananarivo, have revived the weaving of *lamba akotofahana* in recent years. Researching old designs and weaving techniques has allowed them to create high-quality textiles, yet mainly for the export market (see pp.80–1).

Weavers in the town of Arivonimamo use specially adapted looms to reproduce the brightly coloured silk textiles that were once exclusively worn by the Merina nobility.

TRADITION AND CHANGE

It is often suggested that many African textile traditions are in imminent danger of disappearing in the face of mechanization, imported goods, changing social structures and modern fashions. However, while it may be true that certain traditions are in decline, others are taking their place just as rapidly. Throughout the continent there is probably more distinctively African cloth being manufactured today than at any other time.

The use of silk and the traditions it has inspired have always been associated with status and prestige, with aristocracy and royalty, with ancestors and even with deities. Silk textile traditions have been sustained by the continuing demand for prestigious, culturally significant cloths, such as those worn at marriage in North Africa. These glamorous textiles are less exposed to competition from imported goods than more everyday, utilitarian textiles. On the other hand, silk weaving traditions may be more vulnerable to sudden social or political change, as with the disappearance of the *lamba akotofahana* in Madagascar. This type of cloth was directly linked to a royal and aristocratic hierarchy that was largely dismantled during the colonial period.

However, in recent years silk weavers in both Tunisia and Madagascar, while continuing to make silk cloths for local people in styles to which they have become accustomed in the post-colonial era, are also experimenting with designs and patterns not used since the nineteenth century.

In Ethiopia weavers are producing more *shamma* than ever before, and clients can now choose from literally dozens of different designs for the patterned band (*tibeb*). However, the *tibeb* is often woven of rayon rather than the traditional silk – an example perhaps of a modern, synthetic material stimulating rather than stunting an ancient tradition. A similar development can be seen in West Africa, where the silk *kente* cloth of the Asante is now frequently woven in rayon and is worn by a much wider cross-section of Ghanaian society. *Kente* cloths are often worn in place of a suit and tie on formal occasions. The designs and patterns are neither as numerous nor as complex as in the past, but there is no denying that *kente* is now an international phenomenon, familiar around the world.

Not so familiar, but now becoming widely available outside Egypt, are the textiles produced in the old silk weaving town of Naqada in the Nile valley. One design in particular, originally woven in cotton and silk for women in the oases of the Western Desert, has a remarkable story. When demand from their traditional clients waned in the mid-twentieth century, the weavers of Naqada found new markets in Sudan and Libya weaving cloth with the same basic design but using an array of brightly coloured rayon and occasionally lurex yarn (see pp.26–7). When political tension led to the collapse of these new markets, an assisted silk-weaving project was initiated for a new export market. Many women in Naqada began to weave on treadle looms – their use by female weavers had previously been strictly taboo, not just in Egypt but throughout Africa. Today textiles from Naqada may be found in shops around the world. However, those who buy them are probably unaware of the fascinating silk weaving tradition of which they are a part and which encapsulates the dynamism and versatility of so many textile traditions throughout Africa.

A woman using a treadle loom to weave rayon in the old silk weaving town of Naqada in the Nile Valley, Egypt. Traditionally men would have used these looms to weave cotton and silk textiles exclusively for sale to the women of the Bahariya Oasis.

Andalucia
Granada
Algiers
Raf-Raf
Tunis
Mahdia
Azemmour
Fès
MOROCCO
TUNISIA
Gabès
Jerba I.
Mediterranean Sea
Tripoli
Ghadamis Oasis
ALGERIA
LIBYA
Siwa Oasis
EGYPT
Naqada
Nile
Red Sea
Niger
SUDAN
Aksum
Tigray
Gondär
Hausa
Kano
NIGERIA
Addis Ababa
Harär
GHANA
Nupe
Eggan
Kumasi
Yoruba
Asante
Ewe
ETHIOPIA
SOMALIA
KENYA
Pate I.
Lake Victoria
INDIAN OCEAN
ATLANTIC OCEAN
MADAGASCAR
Antananarivo
Merina
Betsileo
national capital
city / town
Ewe people

HEAD SHAWL (*TRUQ'AT*)
Used by women at marriage
ceremonies to cover their head or face.

The black base cloth of this early 20th-century shawl is synthetic. It has densely woven stripes, decorated with tiny embroidered patterns linked to solar motifs. This preoccupation with the sun could be a result of Siwa's earlier link with the ancient Egyptian sun god Amun-Re.
135 × 92 cm (53 × 36¼ in)

THE CHOICE OF COLOURS – PREDOMINANTLY RED, ORANGE AND YELLOW – ARE ASSOCIATED WITH THE SUN. THEIR RESTRICTED USE AGAINST A DARK BACKGROUND GIVES PARTICULAR STRENGTH TO THIS DESIGN.

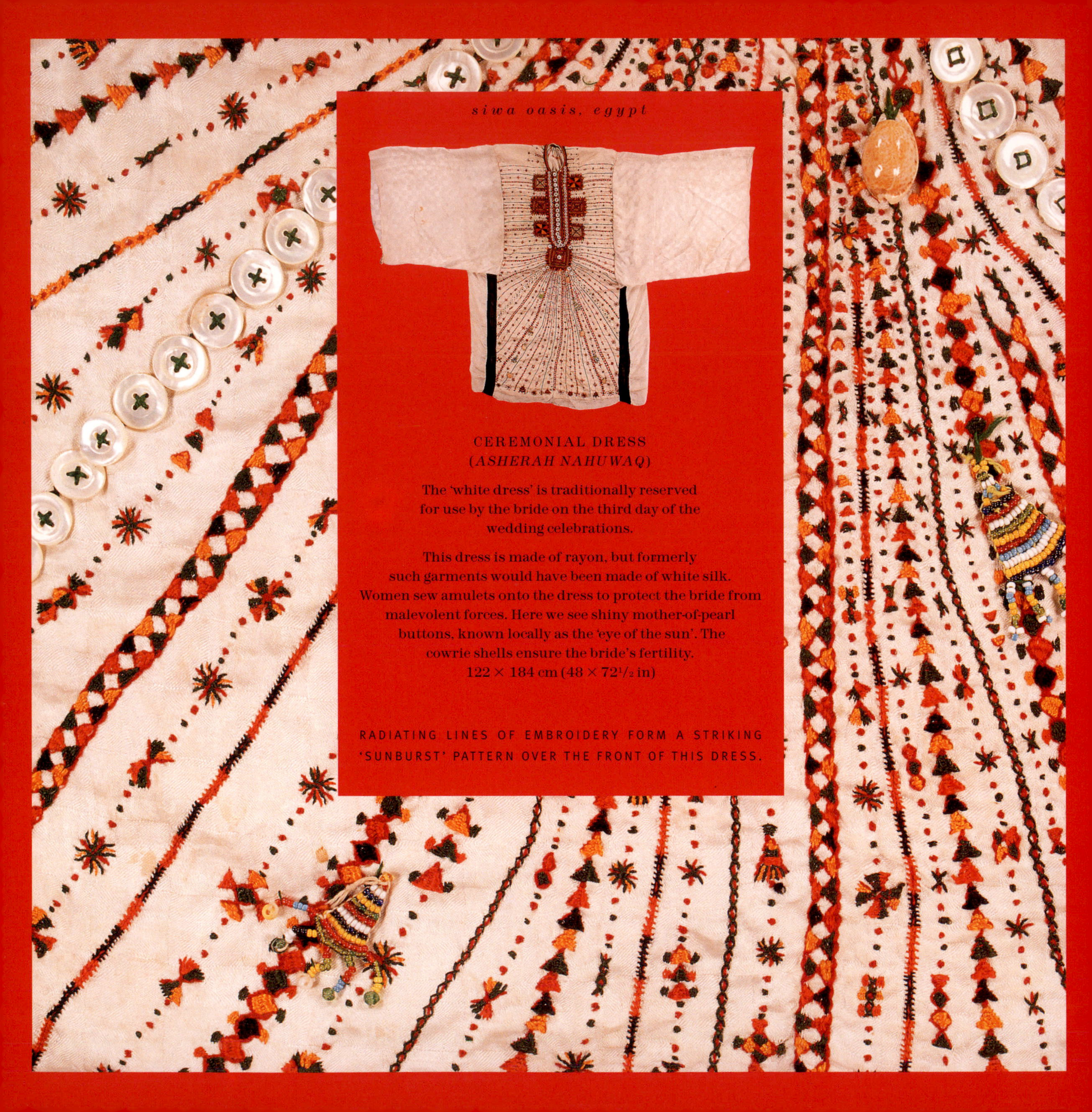

CEREMONIAL DRESS
(*ASHERAH NAHUWAQ*)

The 'white dress' is traditionally reserved for use by the bride on the third day of the wedding celebrations.

This dress is made of rayon, but formerly such garments would have been made of white silk. Women sew amulets onto the dress to protect the bride from malevolent forces. Here we see shiny mother-of-pearl buttons, known locally as the 'eye of the sun'. The cowrie shells ensure the bride's fertility.
122 × 184 cm (48 × 72½ in)

RADIATING LINES OF EMBROIDERY FORM A STRIKING 'SUNBURST' PATTERN OVER THE FRONT OF THIS DRESS.

WOMAN'S MODESTY GARMENT
Women wrap this cloth around the head and body while outside the home.

This tapestry-woven rayon wrap developed from the silk and cotton garments once woven by men for sale to women of Bahariya Oasis. Such cloths are now produced by women using treadle looms. The shimmering cloth conceals the identity of the woman, but artful draping may reveal the body form.
424 × 77 cm (67 × 30¼ in)

THE COMBINATION OF OPPOSING COLOURED THREADS – BLUE-GREEN AND MAGENTA – ENHANCES THE SURFACE SHIMMER EFFECT.

BACK OF MAN'S SILK ROBE (*'ABA'*)
Throughout the Middle East men of all classes wore the '*aba*' – a sleeveless outer robe produced in a wide range of materials.

This fine 19th-century silk robe is deep indigo-dyed with tapestry-woven designs in yellow silk and silver-dipped threads. The armholes are finished with tiny stitches of gold covered cotton thread (below).
135 × 130 cm (53 × 51 in)

DELICATE BANDS OF YELLOW AND SILVER-DIPPED THREAD EXTEND INTO THE SOLID EXPANSES OF INDIGO-DYED CLOTH, CREATING A DAZZLING EFFECT.

THE INSCRIPTION *MA SHA'A ALLAH* (WHAT GOD WILLS) BLENDS HARMONIOUSLY WITH THE ELEGANT DESIGNS COVERING THE BACK OF THE ROBE.

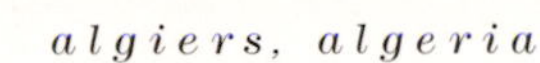

WOMAN'S SHAWL
(*TANSHIFA*)
Reserved for use by
wealthy women at home.

This 19th-century shawl is exquisitely worked, predominantly in violet silk thread, on fine cotton. The principal foliage motifs, worked in Algerian star stitch, are derived from Italian Renaissance forms. Horizontal bands of fine needlework in gold-coloured silk thread separate the bolder elements of decoration.
286×40 cm
($112\frac{1}{2} \times 15\frac{3}{4}$ in)

THE BOLD, VIOLET PATTERNS ARE BEAUTIFULLY BALANCED BY TINY, DELICATE FLORAL MOTIFS SCATTERED THROUGHOUT.

BORDER OF EMBROIDERED CLOTH
Cloth used as a curtain, towel, cushion or mattress cover.

Cloths such as this were embroidered in the home by women and girls. This is a classic example of 19th-century Fès embroidery, with its use of monochrome violet silk thread on a cotton base. The edges of this cloth are rolled and hemmed in blanket stitch.
105 × 80 cm (41 1/3 × 31 1/2 in)

DENSELY EMBROIDERED, LARGELY GEOMETRIC MOTIFS ARE ARRANGED IN FORMAL, REPEATING HORIZONTAL BANDS. THE 'EIGHT-POINTED STAR' OCCURS FREQUENTLY AND THE EMBROIDERY TERMINATES IN ROWS OF STYLIZED TREES (BELOW). SUBTLE COLOUR CHANGES WITHIN THE VIOLET-SILK THREAD ENHANCE THE OVERALL EFFECT.

CEREMONIAL BELT (*HIZAM*)
Such belts were used by women on ceremonial occasions.
They were worn folded in half and wrapped several times around the *qaftan*.

Woven on a draw loom, complex patterns were produced using supplementary heddles. The two sections below show the wide variety of colour and design combinations made possible through using this technique. This is a versatile belt and can be folded to display different patterns for different occasions.
231 × 17 cm (91 × 6¾ in)

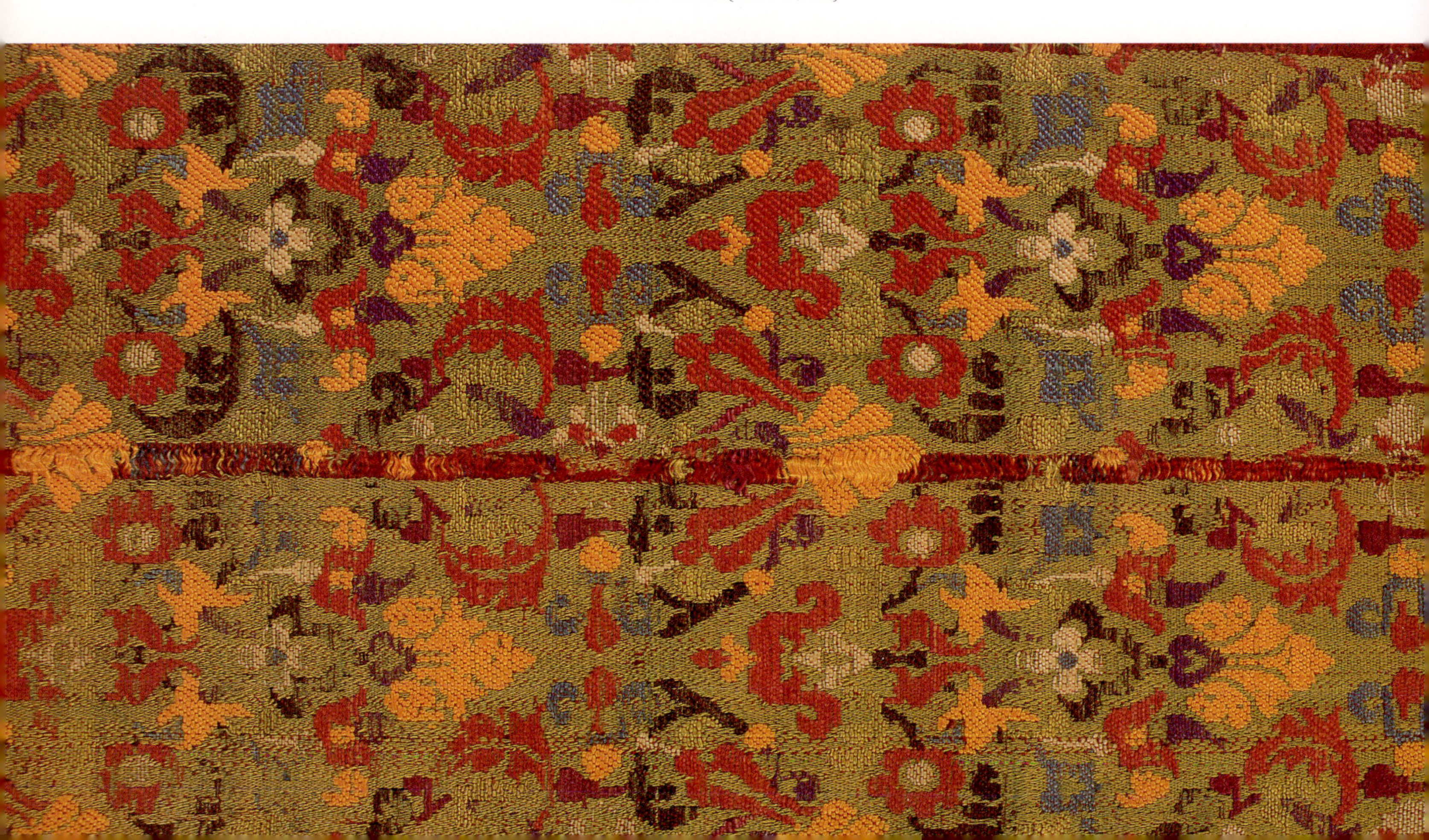

THE GEOMETRIC HERALDIC MOTIFS OF HISPANIC-MORESQUE INSPIRATION CONTRAST WITH THE MORE FLUID, FLORAL MOTIFS INSPIRED BY LATER EUROPEAN AND TURKISH DESIGNS. THE USE OF THE SAME COLOUR PALETTE THROUGHOUT UNITES ALL THE DISPARATE ELEMENTS. THE BASE CLOTH CHANGES FROM BLUE TO GREEN AT THE CENTRE TO PRODUCE TOTALLY DIFFERENT EFFECTS WITHIN THE SAME BELT.

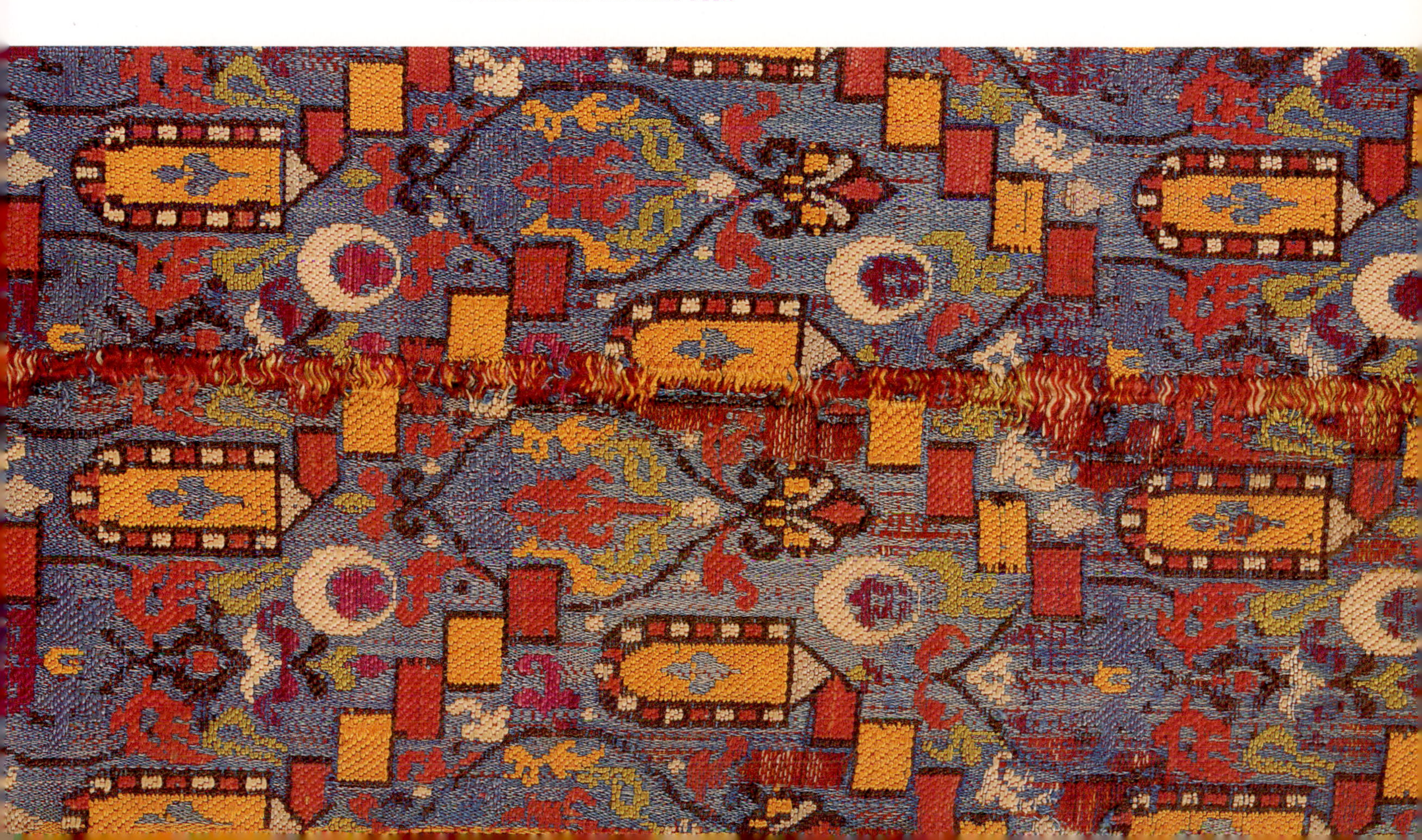

THE FORMAL DESIGNS ON THE FRONT OF THIS BELT PRODUCE A STUNNING ABSTRACT PATTERN ON THE REVERSE. THE IKAT WEAVE IS VISUALLY STRIKING.

CEREMONIAL TUNIC
(*QMAJJA TAWALIY*)
In Mahdia a young girl traditionally prepared twenty–thirty white cotton tunics for her trousseau.

Brides wear such tunics throughout the marriage ceremony. The black silk embroidery (*triza kahla*) on the central panel resembles Andalucian 'blackwork', probably introduced by refugees who settled in Tunisia in the early 16th century. Women make the tassels, which adorn the hem, shoulders and front. The silk ribbons (*hashiya*) that decorate either side of the central plastron are woven by men on draw looms.
79 × 86 cm (31 × 34 in)

THE DESIGN OF THIS TUNIC HAS SEVERAL DISTINCTIVE ELEMENTS: THICK GREEN AND RED SILK TASSELS; MULTI-COLOURED RIBBONS; AND EMBROIDERED EDGE AND CENTRAL PANELS.

THE BOLD, BRIGHT TASSELS AND RIBBONS CONTRAST WITH THE SUBTLE EMBROIDERY. YET ALL THE ELEMENTS ARE SUCCESSFULLY COMBINED BY USING A LIMITED COLOUR PALETTE.

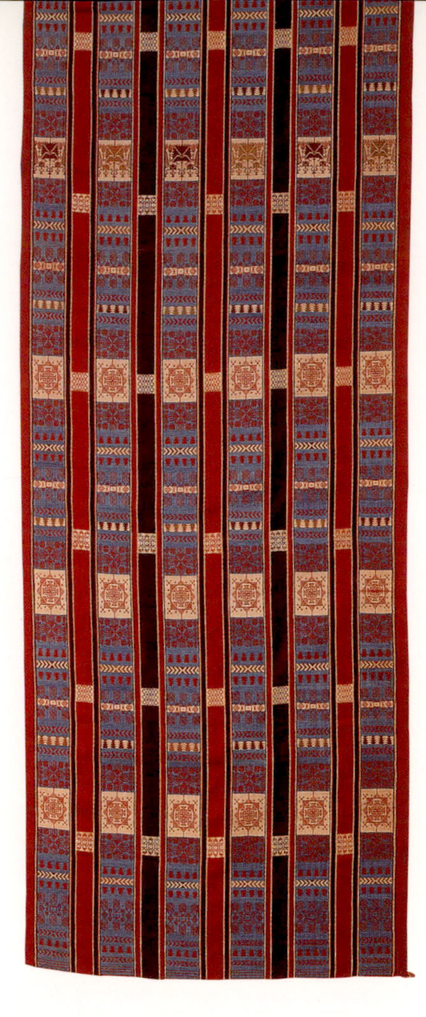

WOMAN'S VEIL (*RIDA'*)
Head-veils were worn by élite women in 19th-century Tunisia when outside the home.

Supplementary heddles are used to produce the complex patterns, which include the 'eight-pointed star' – one of the most common motifs used by Muslims in North Africa. Hand motifs (*khamsa*) of various forms also feature regularly, as well as stylized carnations, derived from Turkey, and cypress trees inspired by the Levant.
278 × 55 cm (109½ × 21½ in)

THIS FORMAL DESIGN IS ACHIEVED USING THREE STYLES OF STRIPES, WHERE THE MOST COMPLEX BECOMES THE DOMINANT FEATURE.

ON A LAVISH BACKGROUND OF DAZZLING TURQUOISE SILK, THIS WAISTCOAT IS DECORATED WITH MOTIFS OF FISH, FLOWERS, STARS, CRESCENTS AND BIRDS THOUGHT TO ENSURE FERTILITY AND TO BRING GOOD LUCK.

THE BACK AND FRONT ARE EQUALLY FLAMBOYANT: 'WINGED' SHOULDERS OF TURKISH INSPIRATION GIVE THE WAISTCOAT ADDED SHAPE; GLITTERING SEQUINS AND INTRICATE GOLD EMBROIDERY ADD RICHNESS.

WOMAN'S WAISTCOAT (*FARMLA*)
Such waistcoats are often worn at marriages and other special occasions.

This 20th-century waistcoat would have been embroidered by the bride or the female members of her family. It has a deeply scooped neckline, designed to reveal the richly decorated plastron of the tunic or blouse worn underneath. The designs are worked in coiled gold and coloured metal thread and sequins. Tiny coral beads are sewn to the front of the waistcoat; in the Mediterranean coral is associated with fertility.
40 × 45 cm (15¾ × 17¾ in)

WOMEN'S SHOES (*TARKASIN*)
These slippers formed part of the trousseau of a bride from Ghadamis Oasis.

The shoes are hand-stitched. Their red-dyed leather uppers are richly decorated with multicoloured silk thread embroidery. The tongues are cut into the form of a hand. Hand motifs, sometimes described as the 'hand of Fatima', are used throughout the Islamic world to protect against the evil eye. The number five is also believed to be propitious.
25 × 11 × 13 cm (9¾ × 4⅓ × 5 in)

THESE SHOES ARE LIBERALLY ENDOWED WITH PROTECTIVE DEVICES. METAL TACKS ARE USED EXTENSIVELY BECAUSE SHINY OBJECTS ARE BELIEVED TO REFLECT THE EVIL EYE.

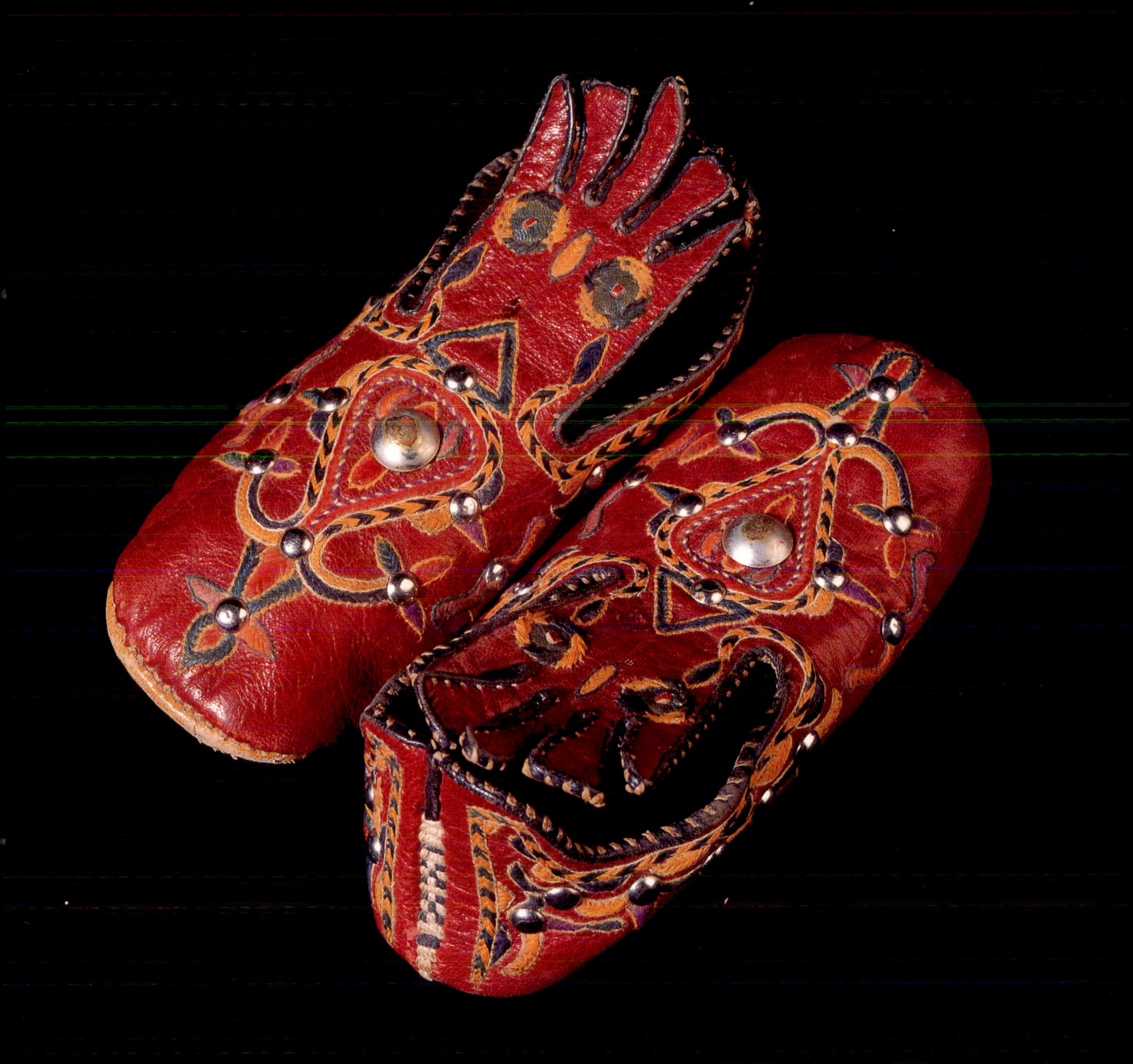

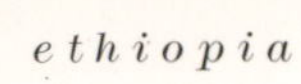

PATTERNED
BAND (*TIBEB*)
The *tibeb* is woven into
the finest *shamma*, a shawl made
of several layers of cotton cloth
sewn together.

The *shamma* is worn
by men or women throughout
Ethiopia. It is woven by men using
double-heddle treadle looms, and the
vibrant design is achieved by inserting
numerous supplementary shed
sticks into the warp. The width and
complexity of this *tibeb* show the
wearer was a person of high status.
Shamma: 235 × 149 cm
($92\frac{1}{2} \times 58\frac{1}{4}$ in)

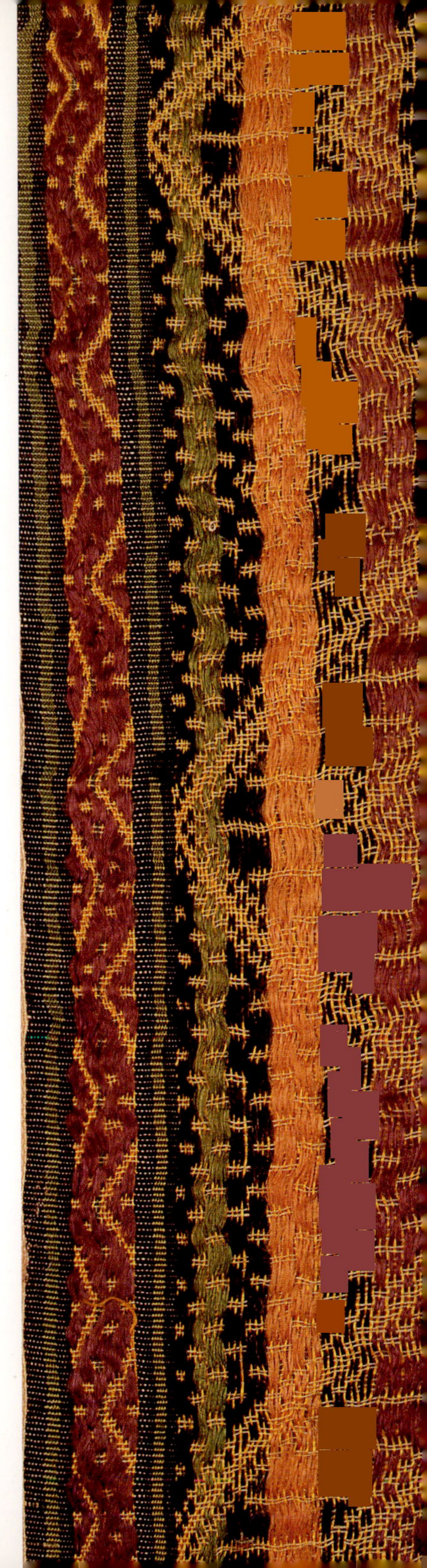

A MUTED RANGE OF TERTIARY
COLOURS ENSURES THAT THIS
COMPLEX GEOMETRIC DESIGN
POSSESSES A VISUAL UNITY.

THE STRIKING GRAPHIC FIGURES REPRESENT KING BAKAFFA'S BODY ON A FUNERAL BIER, FLANKED BY ANGELS WITH STYLIZED WINGS (LEFT). THE DOMINANT RED AND ORANGE COLOURING HAS BEEN ENLIVENED BY THE INTRODUCTION OF BLUE AND GREEN LINES.

HANGING SCREEN
This 18th-century textile from Ethiopia was once part of a triptych, designed to separate the inner sanctum from the main body of a church.

This is the largest tablet-woven textile in the world and is woven entirely of imported Chinese silk on a tablet-weaving loom. The design probably represents a funerary procession with figures arranged in descending hierarchy: at the top is the body of King Bakaffa, followed by his wife Queen Mentuab, his son Iyasu, and then clerics, governors and soldiers.
306 × 63 cm (20 1/2 × 24 3/4 in)

NOBLEMAN'S CAPE *(LAMD)*

The nobleman's cape developed from the earlier honorific lion-skin capes awarded to successful soldiers and hunters by the Ethiopian emperor.

This sumptuous garment is made of purple velvet with an imported silk lining. The velvet is covered in clusters of silk embroidery, particular to this commission. Delicate leaf-like motifs combine with more complex designs, braided with precious metal. The larger pieces of metal are sewn on individually. A silver-gilt clasp draws the cloak together.
99 × 86 cm (39 × 33³/₄ in)

THE ORNATE, ARABESQUE PENDENT PANELS REPRESENT, IN STYLIZED FORM, THE LIMBS AND PAWS OF A LION. THE DRAMATIC DESIGN IS ACHIEVED BY COMBINING GOLD AND BRIGHT COLOURS WITHIN BOLD SHAPES THAT CONTRAST WITH THE DARK GROUND.

EMBROIDERED TUNIC AND TROUSERS
This style of costume was worn by Ethiopian noblewomen of the central and northern provinces during the 19th century.

Both tunic and trousers were embroidered by the women themselves, using imported Chinese silk on a base cloth of Manchester calico. Its lively design has kept its appeal: very similar patterns are now immensely popular on printed fabrics throughout eastern Africa.
107 × 165 cm (42 × 65 in)

THE OPEN-FORKED DESIGN SURROUNDING THE DENSELY GEOMETRIC CENTRAL PANEL GIVES A FIERY EFFECT TO THIS OTHERWISE PLAIN GARMENT.

THE CRUCIFORM MOTIF IS IN THE FORM OF ONE OF THE MANY PROCESSIONAL CROSSES USED BY THE ETHIOPIAN ORTHODOX CHURCH.

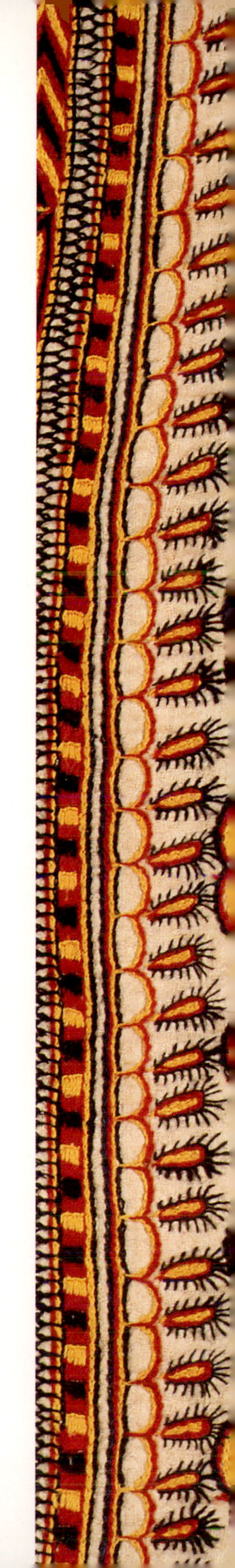

WOMAN'S DRESS (*KAMIS*)
This style of dress developed from the 19th-century silk-embroidered tunic worn by noblewomen on special occasions.

It is made of Ethiopian cotton, hand-woven on a treadle loom. The fine white muslin is bunched at the waist, causing it to hang in numerous folds. The borders are woven and the central panel is embroidered by men, using rayon and lurex. The crucifix is similar to those that appear on Medieval Christian manuscripts from Ethiopia.
140 × 117 cm (55 × 46 in)

THE PLAIN WHITE CLOTH CONTRASTS WITH THE BOLD COLOURS AND DESIGNS OF THE COMPLEX EMBROIDERED CROSS AND THE DENSELY WOVEN BORDERS, DECORATED WITH AN ANCIENT CHEVRON PATTERN.

SILK CLOTH

This cloth was described by the collector, Thomas Bowdich, as a girdle manufactured in 'Houssa' (Hausa), northern Nigeria.

One of the earliest cloths in the Museum's collection, this cloth was woven in a single wide strip. It is composed of simple warp stripes using yellow, white, blue and magenta waste silk (*alharini*).
190 × 30 cm
(74¾ × 11¾ in)

THIS DELIGHTFUL DESIGN IS MADE UP OF FINE BANDS OF YELLOW, GREEN, BLUE AND WHITE STRIPES BALANCED WITH BANDS OF SOLID RED.

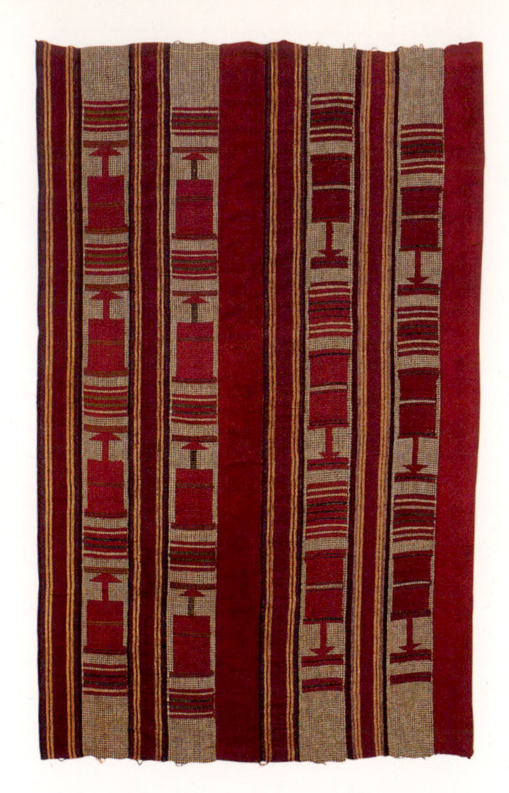

WOMAN'S CLOTH (*GANDERE*)
The use of magenta silk (*alharini*) in this 19th-century cloth indicates it would have been worn for special occasions.

This cloth is made of ten narrow strips, woven separately and then sewn together. Strips of checked indigo-dyed and white-checked cotton alternate with strips using imported *alharini*. The one-sided supplementary weft floats, also using *alharini*, are woven into the blue and white check; they are characteristic of Yoruba weaving and an Ilorin speciality.
180 × 111 cm (51 × 43 3/4 in)

THE GEOMETRIC INLAY PATTERNS FEATURE *QUR'AN* BOARDS (WOODEN WRITING BOARDS). THEIR HANDLES LOOK LIKE ARROWS AND CREATE A SENSE OF MOVEMENT SINCE THE ARROWS ON THE LEFT AND RIGHT COLUMNS FACE IN DIFFERENT DIRECTIONS. THE DRAMATIC COMBINATION OF GREEN AND DEEP PINK ENHANCES THE DESIGN.

SILK AND COTTON WRAP

Worn by men, draped around the body and over the left shoulder and arm.

This 20th-century cloth, made by the Asante people of Ghana, is composed of twenty-four narrow strips woven by men using a double-heddle loom. The strips are sewn together to produce a magnificent overall effect. Bold warp stripes in yellow, green, red and blue combine with weft-float patterns.

306 × 183 cm (104½ × 72 in)

A VARIETY OF WEFT-FLOAT DESIGNS ALTERNATE WITH THE SIMPLE WARP STRIPES TO CREATE A REGULARLY SPACED AND BALANCED COMPOSITION.

THE BLUE IN THE MAIN DESIGN IS HIDDEN IN THE BORDER, WITH ITS DISTINCTIVE BUT COMPARITIVELY SIMPLE ALTERNATING PATTERN.

THIS IS AN INTENSELY PATTERNED, DYNAMIC TEXTILE WHICH JUXTAPOSES COLOUR AND DESIGN.

THE NON-ALIGNMENT OF THE PATTERN ELEMENTS ADDS TO THE SENSE OF MOTION AND VIBRANCY.

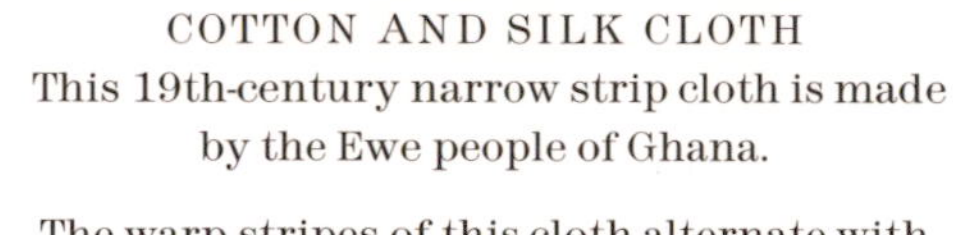

COTTON AND SILK CLOTH
This 19th-century narrow strip cloth is made by the Ewe people of Ghana.

The warp stripes of this cloth alternate with weft-faced blocks in machine-spun cotton, embellished with supplementary weft floats in silk. There are seven different patterns of warp stripes as well as numerous weft-faced designs. It uses ready-dyed machine-spun cotton, which provides a wider range of colours than cotton that has been locally spun and dyed.
278 × 189 cm (109 1/2 × 74 1/2 in)

MAN'S ROBE (*RIGAN GIWA*)
An exquisite 19th-century robe of honour.

The gown's narrow strips are half patterned and half solid colour; these strips are folded and sewn so that the solid colour forms an inner lining. It is made with a cotton and imported magenta waste silk warp and a cotton weft. The embroidery is in imported silk thread. The cost and physical weight of this garment make it a prestige item.
138 × 252 cm ($54^{1}/_{3} \times 99^{1}/_{4}$ in)

THE MAIN DESIGN ON THIS ROBE IS ASYMMETRICAL. A POPULAR 'EIGHT KNIVES' PATTERN WITH LONG TRIANGULAR ELEMENTS IS ARRANGED IN TWO GROUPS OF THREE AND TWO. THIS PATTERN GUARDS AGAINST THE EVIL EYE.

STYLIZED VERSIONS OF THE 'EIGHT-POINTED STAR' AND THE 'HOUSE OF FIVE', NINE INTERLOCKING SQUARES OF WHICH FIVE ARE EMBROIDERED, GUARD AGAINST THE EVIL EYE. THE CROSSED CIRCLE WITH SPIRAL IS REFERRED TO AS *TAMBARI* ('KING'S DRUM'); IT IS ALSO PROTECTIVE AND IN HAUSA IS ASSOCIATED WITH CHIEFTAINCY.

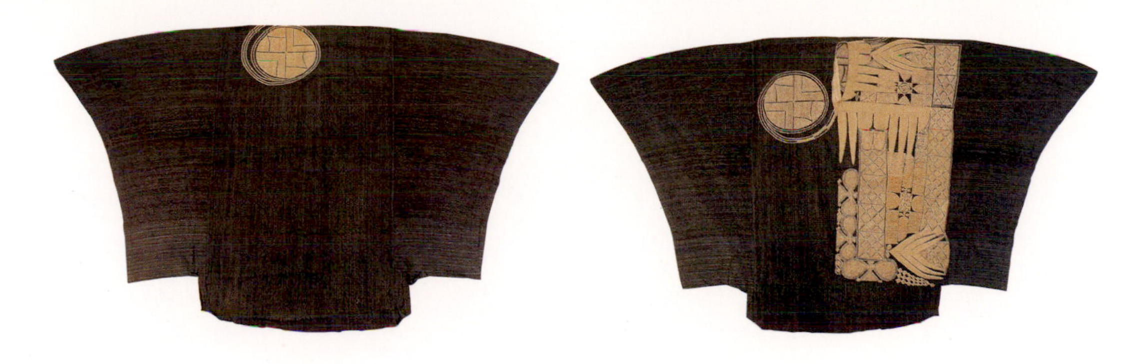

MAN'S ROBE

This prestigious robe would have been worn on special occasions.

It is woven in narrow strips using white and indigo-dyed cotton to produce a distinctive pattern known as the 'guineafowl' design. It is embroidered using imported silk thread. The right-hand side of the front of this robe is dominated by an expanded form of the 'eight-knives' pattern. Its intricate embroidery extends almost to the hem of the garment.
148 × 260 cm (58¼ × 102½ in)

THE 'TRIPLE BARB' (*GABIYA*) PATTERN IS CHARACTERISTIC OF NUPE EMBROIDERY AND IS PARTICULARLY ASSOCIATED WITH THE TOWN OF BIDA.

THE ELEGANT CROSSED CIRCLE AND SPIRAL MOTIF IS FOUND ON THE BACK AND THE RIGHT CHEST OF THE ROBE, AS ON ALL NORTHERN NIGERIAN GOWNS.

CEREMONIAL CLOTH (*ITAGBE*)
Women produce these ceremonial *itagbe* cloths which are worn by chiefs of the Ogboni society.

The cotton base cloth is decorated with silk. The slits at either end of this cloth are woven in using an extra set of weft threads. Colourful bands with silk inlay pattern are separated by vivid tufts of silk. The tufts have protective qualities. Similar tufts are seen on cloths associated with herbal doctors or used as baby-carriers.
151 × 42 cm ($59^1/_2$ × $16^1/_2$ in)

THIS BOLD, TACTILE DESIGN USES VIBRANT COLOURS. IT IS ORGANIZED INTO TUFTED SECTIONS (*SHAKI*) SEPARATED BY STEPPED GEOMETRIC

OUTER PANELS AND A RHYTHMIC CENTRAL PANEL, DECORATED WITH THE 'CROCODILE' PATTERN, (*ONI*).

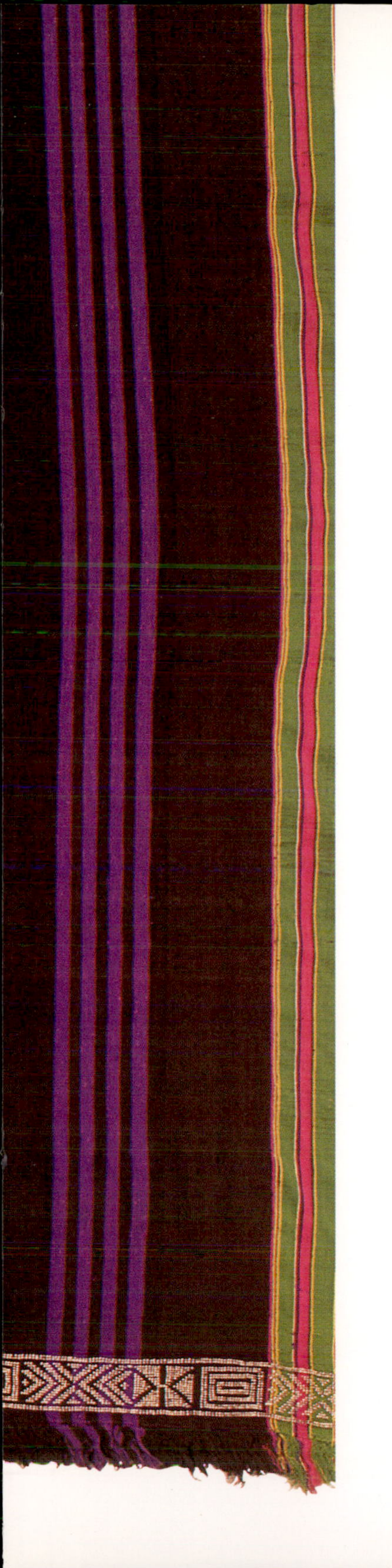

SILK TEXTILE (*LAMBA*)
Such Malagasy textiles may be used as a shawl or a loincloth, and fine, prestigious examples such as this may also be used as a burial shroud.

This textile is woven by the Betsileo people in three separate strips sewn together. The decorative border is created by clipping metal beads onto the warp threads in a particular order so that they appear as a pattern when woven in with the weft.
228 × 60 cm (89¾ × 23½ in)

THIS SIMPLE, YET EFFECTIVE, DESIGN HAS BEEN CREATED BY USING COLOUR OPPOSITES. BANDS OF OPPOSING RED AND GREEN STRIPES DIVIDE THE BLACK AND PURPLE STRIPES.

SILK CLOTH
(*LAMBA AKOTOFAHANA*)
Such Malagasy cloths were used on important occasions and were sometimes used as burial shrouds (*lamba mena*).

This cloth was woven on a fixed, single heddle loom. Supplementary heddles suspended over the loom create the bands of complex pattern that float across the warp. Blue-green was a colour exclusive to the royal clan.
244 × 163 cm (96 × 64 in)

DECORATIVE BANDS ENLIVEN THIS STRIPED TEXTILE. THE PREDOMINANT PATTERNS ARE THOUGHT BY SOME TO BE VARIATIONS OF THE 'TREE OF LIFE' MOTIF.

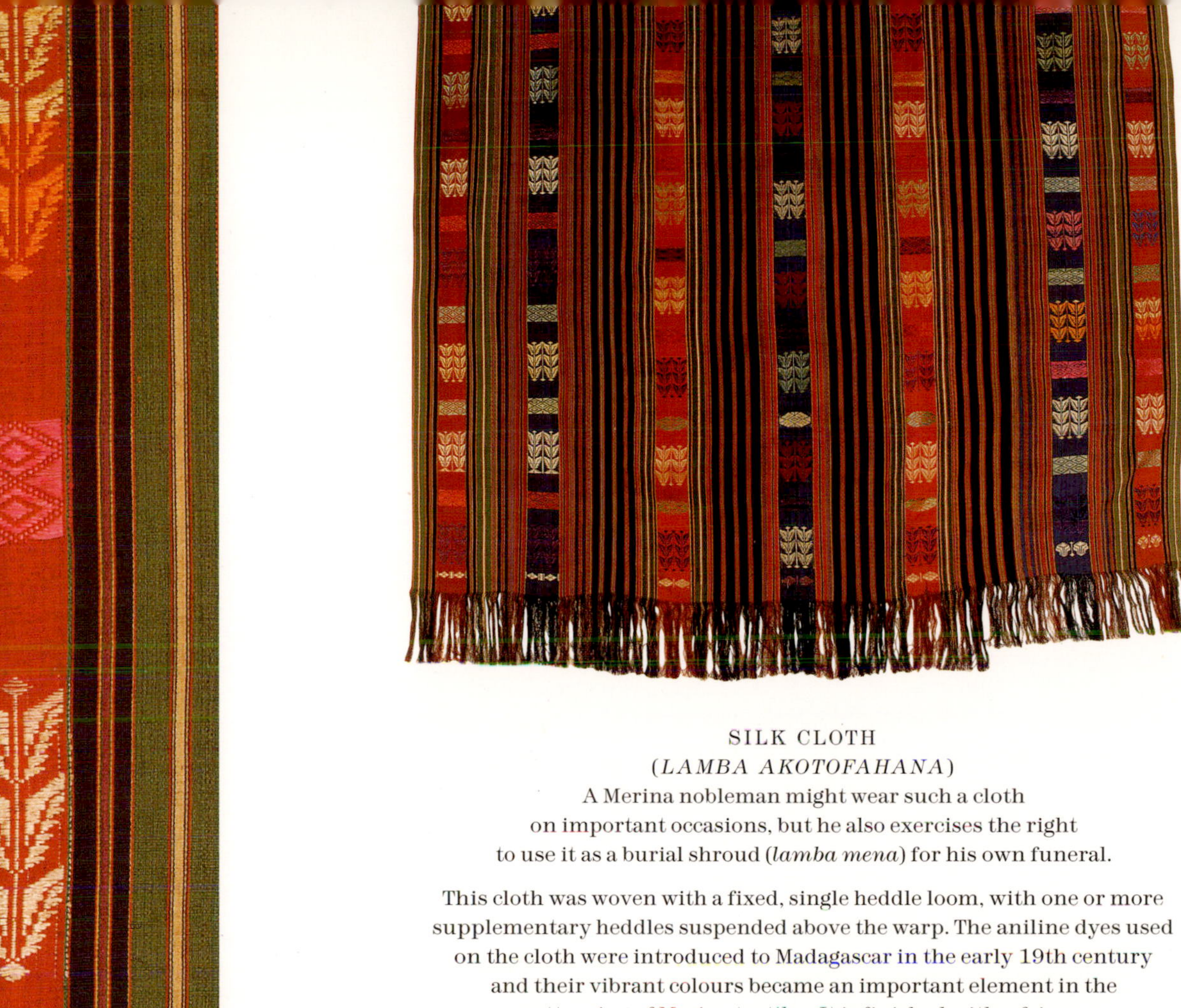

SILK CLOTH
(*LAMBA AKOTOFAHANA*)
A Merina nobleman might wear such a cloth on important occasions, but he also exercises the right to use it as a burial shroud (*lamba mena*) for his own funeral.

This cloth was woven with a fixed, single heddle loom, with one or more supplementary heddles suspended above the warp. The aniline dyes used on the cloth were introduced to Madagascar in the early 19th century and their vibrant colours became an important element in the patterning of Merina textiles. It is finished with a fringe.
244 × 163 cm (96 × 64 in)

IN COMPARISON TO THE TEXTILE ON THE PREVIOUS TWO PAGES, THIS CLOTH USES A MUCH BROADER AND BRIGHTER COLOUR RANGE.

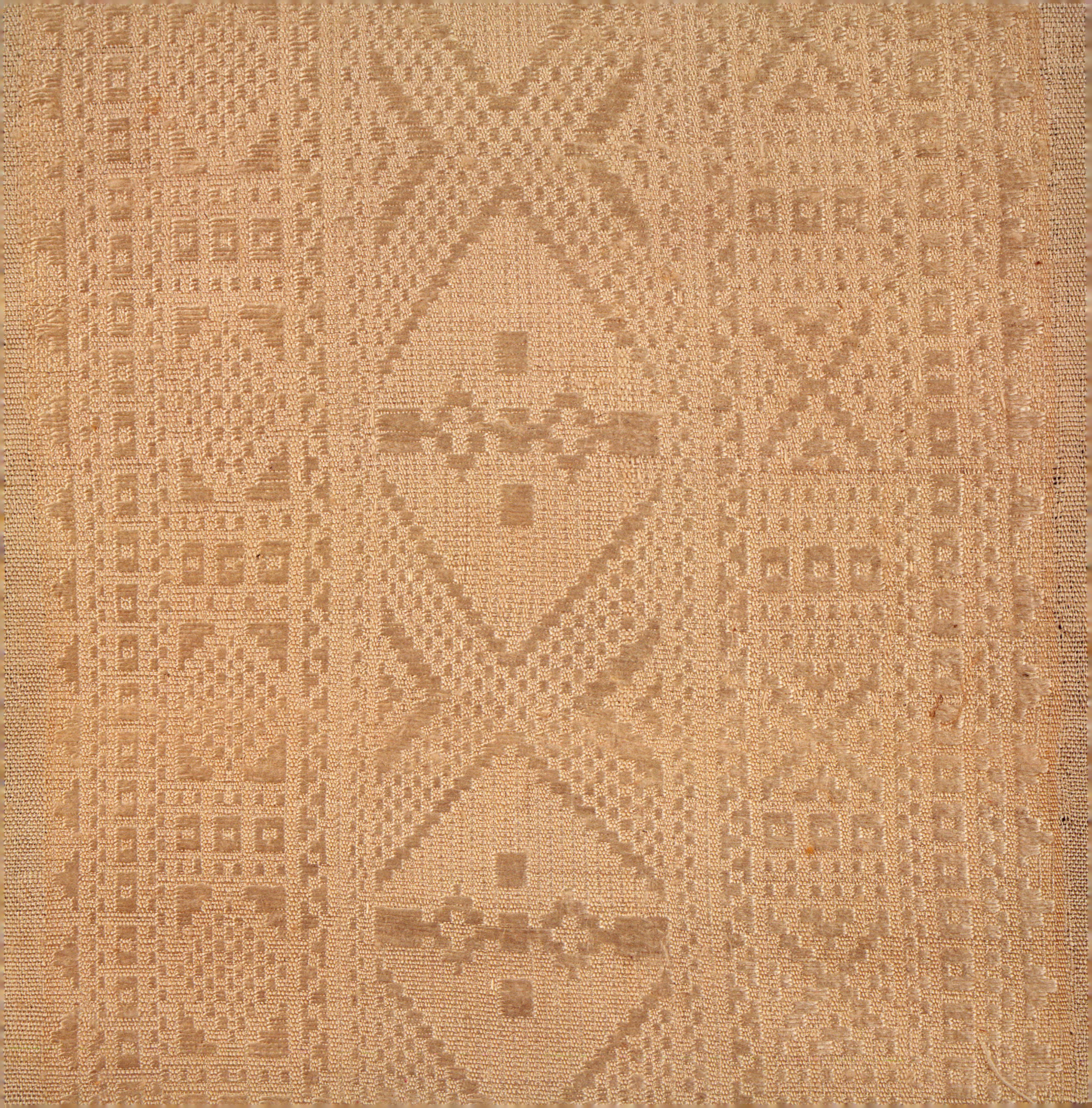

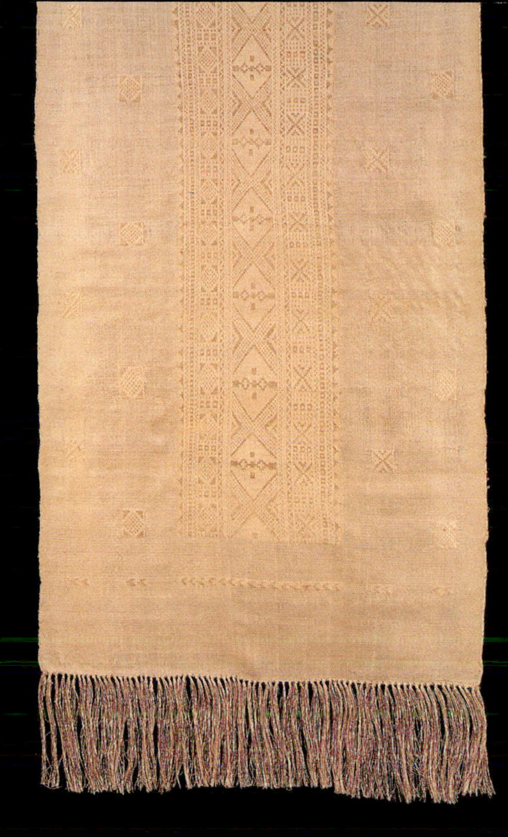

SILK CLOTH (*LAMBA*).
In the early 20th century, high-ranking Merina began to wear a white silk *lamba* as a sign of status.

This *lamba* incorporates some of the traditional motifs of the 19th-century multicoloured *lamba akotofahana* and is woven on the same type of loom. However, its subtle white-on-white design is a masterpiece of understatement.
228 × 60 cm (89¾ × 23⅔ in)

THE CENTRAL PANEL OF THIS PLEASING TEXTILE IS RICHLY DECORATED WITH SUBTLE COMBINATIONS OF WEFT-FLOAT DESIGNS.

SILK TEXTILE (*LAMBA*)
Textile woven on a specially modified loom for the export market.

The weaver of this modern cloth is inspired by the 19th-century patterns and designs of the *lamba akotofahana* of pre-colonial Madagascar. It is not an exact replica, but its making requires the same high levels of craftsmanship and expertise as earlier textiles demanded.
232 × 57 cm (91 1/3 × 22 1/2 in)

THIS IS ANOTHER EXAMPLE OF THE STRIPED MALAGASY *LAMBA*; HOWEVER, BY USING A SATISFYINGLY RESTRAINED RANGE OF COLOURS, THE WEAVER PRODUCES A VERY DIFFERENT EFFECT.

glossary

'*aba*' man's sleeveless outer coat worn throughout the Middle East.

adanudo one of the main cloths woven by the Ewe peoples for their own use. Often made of silk with numerous WEFT-INLAY designs.

alharini magenta WASTE SILK originally imported from Tunisia, later from Europe. Traded to Kano via trans-Saharan trade routes since the eleventh century.

Anaphe infracta a species of silkworm found wild in parts of Africa that is communal by nature and produces clusters of cocoons encased in a brown silk outer pod. Silk is obtained by de-gumming and spinning.

Anaphe moloneyi from the same genus as above. The silkworms form individual cocoons without an outer casing. The spun silk is used primarily for embroidery.

aniline dyes synthetic dyes imported into Africa from the early nineteenth century and extensively used by silk weavers.

asaia royal cloths woven exclusively for the Asantehene (Ghana) and family.

beniqa Algerian woman's embroidered linen cap used to wrap up damp hair after bathing.

biskri wrap-around woman's garment woven on Jerba island, Tunisia, characterized by its elaborately decorated end panels.

Bombyx mori domesticated silk moth that feeds primarily on mulberry leaves. It is reared extensively in Europe and China producing silk of outstanding quality and quantity. Its intensive domestication has resulted in it being rendered blind, flightless and without digestive organs.

draw loom loom operated by the weaver and one or more assistants who 'draw up' the supplementary HEDDLES to form the pattern elements on a textile.

famadihana second-burial ritual in Madagascar when the remains of the dead are disinterred and wrapped in a new silk shroud.

flying shuttle invented by John Kay in the early eighteenth century. A modified SHUTTLE BOAT runs along a track beneath the warp threads and is activated by the weaver pulling on a cord. This device increased output and allowed broader cloths to be made by a single weaver.

hammam steam bath.

hashiya silk ribbons woven on a DRAW LOOM in Mahdia. Used to decorate a variety of costumes and furnishings.

heddle rod to which WARP threads are attached by means of loops. When raised it forms the SHED.

double two heddles attached by means of loops to both groups of WARP threads.

supplementary more than two heddles; used in raw looms to create complex patterns.

hizam belt used by men and women.

horizontal ground loom loom with warped threads mounted horizontally; warp beams are often pegged to the ground.

hram nfasiy wrap-around ceremonial cloth worn for the first time at marriage in El-Jem, Tunisia.

kamis woman's dress in contemporary Ethiopia, the name deriving from the French '*chemise*'.

kente commonly used as a general term for Asante silk textiles.

lamba generic name for cloth in Madagascar.

lamba akotofahana silk cloth of complex pattern used by the nobility of the nineteenth-century kingdom of Imerina in Madagascar.

lamd honorific cape worn by high-ranking secular and religious officials in Ethiopia.

matab plaited band of blue silk worn around the neck in Ethiopia as a symbol of the Christian faith.

nsaduaso fine silk cloth with WEFT-FLOAT patterns woven by the Asante, Ghana.

raw silk silk retaining its natural gum that resists dye. Usually de-gummed by heating in an alkaline solution.

rida' ahmar literally 'red outer garment' worn by women in Tunisia on ceremonial occasions.

shafts slats of wood on a TREADLE LOOM that support the HEDDLES.

shamma toga-like shawl worn by men and women throughout Ethiopia.

shed the opening made between the WARP threads through which the WEFT is passed during weaving.

shed stick sticks placed between WARP elements to allow the WEFT to be passed through to form patterns.

shuttle boat device used to pass thread through the SHED.

silk natural filament obtained from the cocoons of silkworms.

tanshifa Algerian woman's embroidered linen shawl; reserved for special occasions.

tapestry WEFT-faced fabric built up of many areas of colour to form pattern.

tibeb silk band of complex pattern woven into the more prestigious types of SHAMMA in Ethiopia.

tiraz Persian term meaning 'embroidery'.

treadle loom sturdy foot-operated floor loom with SHAFTS for raising and lowering threads. The shafts are attached to treadles which increase the speed of weaving.

warp longitudinal threads in a woven structure.

waste silk by-product of silk processing; obtained from broken cocoons which cannot be reeled (wound), the short threads are spun in a manner similar to cotton and wool.

weft transverse threads that interlace with the WARP.

weft float supplementary weft that is 'floated' across a warp-faced textile to form patterns.

weft inlay supplementary discontiuous WEFT forming blocks of inlay pattern.

selected reading

General

Picton, J. and J. Mack, 1989, *African Textiles*, (2nd ed.), London.

Schaedler, K-F., 1987, *Weaving in Africa South of the Sahara*, Munich.

Spring, C, 1997, *African Textiles*, (2nd ed.), London.

North Africa

Bairam, A., 1984, 'Le tissage de la soie à Tunis à la veille de l'indépendence,' *Cahiers des Arts et Traditions Populaires,* no. VIII:41–52.

Ben Tanfous, A., 1996, 'Relation vestimentaire entre Jerba et Tripoli,' *Cahiers des Arts et Traditions Populaires,* no. XI:33–40.

Golvin, L., 1950, 'Le "métier à la tire." Des Fabricants de Brocarts de Fès,' *Hespéris*, XXXVII:21–52.

Serjeant, R.B., 1951, 'Material for a History of Islamic Textiles up to the Mongol Conquest: The Maghreb,' *Arts Islamica*, vol. XV–XVI:41–54.

Spring, C. and J. Hudson, 1995, *North African Textiles*, London.

Stone, C., 1985, *The Embroideries of North Africa*, London.

Vivier, M-F., 1996, 'Les broderies marocaines,' *in De soie et d'or: broderies du Maghreb*, Institut du Monde Arabe, Paris.

Ethiopia

Balicka-Witakowska, E. and M. Gervers, 1996, 'Monumental Ethiopian tablet-woven silk curtains: a case for royal patronage,' *Burlington Magazine,* (June), pp.375–85.

Brown, H., 1988, 'Siyu: Town of the Craftsmen,' *Azania*, vol. XXIII:101–113.

Gervers, M., 1995, 'The Death of King Bakaffa: a Story told in Silk,' *Rotunda,* 27:4, pp.34–9.

Girma, F., 2000, 'Cape (lemd),' in J. Mack (ed.), *Africa: Arts and Cultures,* London.

Messing, S.D., 1960, 'The Nonverbal Language of the Ethiopian Toga,' *Anthropos,* 55:3–4, pp.558–60.

Lefebvre, T., 1845–8, *Voyage en Abyssinie,* Paris.

Pankhurst, R., 1982, 'History of Ethiopian Towns: from the Middle Ages to the early nineteenth century,' *Äthiopistische Forschungen,* vol. 8:261, 298–302.

Pankhurst, R., 1990, *A Social History of Ethiopia,* Addis Ababa.

West Africa

Cole, H.M. and D.H. Ross., 1977, *The Arts of Ghana,* Los Angeles.

Ene, C.J., 1964, 'Indigenous Silk-weaving in Nigeria,' *Nigeria Magazine,* no. 81 (June), pp.127–37.

Gilfoy, P.S., 1987, *Patterns of Life: West African Strip-Weaving Traditions,* Washington DC.

Heathcote, D., 1976, 'Hausa Embroidered Dress,' *African Arts,* vol. V, 2:12–19.

Johnson, M., 1973, 'Cloth on the banks of the Niger,' *Journal of the Historical Society of Nigeria,* vol. 6, no. 4 (June), pp.353–63.

Kriger, C., 1988, 'Robes of the Sokoto Caliphate,' *African Arts,* vol. XXI, 3:52–7, 85.

Lamb, V., 1975, *West African Weaving*, London.

Lamb, V. and J. Holmes, 1980, *Nigerian Weaving,* Roxford.

McLeod, M.D., 1981, *The Asante,* London.

Perani, J., 1979, 'Nupe Costume Crafts,' *African Arts*, vol. XII, 3:52–7, 96.

Madagascar

Mack, J., 1986, *Madagascar: Island of the Ancestors,* London.

Mack, J., 1989, *Malagasy Textiles,* Aylesbury.

Peers, S., 1997, 'Malagasy Lamba: Silk weaving among the Merina of Madagascar,' *Hali*, issue 95 (Nov), pp.82–5.

Peers, S., 1995, 'Weaving in Madagascar,' in J. Picton (ed.), *The Art of African Textiles: Technology, Tradition and Lurex*, London.

Rasoamampionona, C., 2000, 'Textile (lamba akotofahana),' in J. Mack (ed.), *Africa: Arts and Cultures,* London.

museum accession numbers

PAGE	ACC. NO.
2	1868,10–1.22
4	1934, 3–7.189
6	2000 Af 5.17 a + b
7	1968 Af 13.1 (top)
7	1934 3–7.198 (left)
7	1993 Af 16.1a (right)
7	1951 Af 12.230 (below)
22	1991 Af 11.4
24	1991 Af 11.1
26	1993 Af 24.7
28	1994 Af 3.1
30	1907,10–16.11
32	1916,8–5.1
34	1970 Af 2.2
38	1987 Af 1.78
40	1907,10–16.9
42	1998 Af 1.113
44	1981 Af 5.2 a + b
46	1974 Af 11.5
48	1968.10–1.22
50	1974 Af 11.10
52	AB1
54	1993 Af 16.1a
56	1881,11–14.25
58	1900–36
60	1934 3–7.198
62	1934.3–7.165
64	1920, 2–11.1
68	1934, 3–7.215
70	1951 Af 12.230
72	1968 Af 13.1
76	1988 Af 15.1
78	1985 Af 17.1
81	1993 Af 14.7
INSIDE COVER	1951 AF 12.230

publisher's acknowledgements

The textiles featured in this book are drawn from the collections of the British Museum's Department of Ethnography and have been selected from the viewpoint of their cultural significance as well as their design and technical merit.

We should like to express our thanks to the many people who have helped us in the production of this book, and in particular from the Museum staff: Helen Wolfe, Kate Johnson and Heidi Cutts. Paul Welti, the art director, must be credited not only for his arresting juxtaposition of illustrations and text, but also for his contribution to the captions analyzing the designs.

picture credits

All pictures are the copyright of The Trustees of The British Museum with the exception of the following:
p.9 J. Mack; p.10 Chris Spring; p.11 *Musée National des Arts et Traditions Populaires,* Algiers; p.13 Stephen Bell; p.15 J. Picton; p.19 Simon Peers; p.20 J. Mellors

authors' acknowedgement

The authors would like to thank Mike Cobb and Heidi Cutts for their support and enthusiasm, and thanks to Paul Welti for his contribution to the design captions.

index